CAREERS WITH GEOGRAPHIC INFORMATION SYSTEMS (GIS)

YOU LIVE IN AN AGE IN WHICH OLD careers disappear and are replaced by new ones on a constant basis. Have you used a printed paper map lately? They are rarely used today, because the paper-and-ink cartography of the past has given way to the geographic information systems of today. Often referred to simply as GIS, GIS applications, or GIS platforms, geographic information systems enable us to find our way around the modern world. From the Google maps in our smartphones to the sophisticated targeting systems used by the military, GIS platforms have become a common part of everyday life.

That means there are many opportunities for dedicated professionals to take GIS to the next level. GIS platforms are used by geographers, scientists, real estate agents, property developers, government agencies, conservationists, in military specialties, by individuals driving their cars, and in many other useful ways.

These users may work with the scientists and computer programmers who design GIS platforms. The purpose of any GIS platform is to work well in the field, making constant collaboration with end-users extremely important. Most GIS systems, in fact, are customized to achieve a very specific purpose. Some businesses maintain in-house GIS departments to provide constant geographic data, while others contract with dedicated GIS providers whenever they need geographic data.

This report will give you information on how to prepare for your career, what kind of academic and vocational training you will need, how much money you can expect to earn at various points in your career, and even what you may like or dislike about the work. You will also find sections on where to look for work, the state of the job market, and how to get some experience sooner than you might think. Be sure to check out the links on the last page of this report.

WHAT YOU CAN DO NOW

THERE ARE WAYS TO GET SOME EXPERIENCE with geographic information systems while you are still in school.

You may think of GIS as a big-budget technology used only by businesses and government agencies, but that is not true. The common navigation systems used in cars and smartphones use the same basic Global Positioning System technology found in the most advanced GIS products. GIS is a broad term used to describe any system that allows a user to visualize, interpret and analyze geographic data. Systems used in extremely sophisticated applications, like military uses, can gather billions of individual data points that can then be visualized and analyzed in myriad ways. The mapping program in the smartphone in your pocket allows you to visualize, interpret and analyze simple geographic data in the form of a two-dimensional map to help you get from here to there. That makes it a GIS device. You have access to others, like Google Maps and MapQuest.

Get involved in orienteering, a mentally and physically challenging sport that will put your GIS skills to the test. Orienteering is the original GIS challenge. Teams or individuals are provided with a series of very specific points on a map and are required to find their way to each one while racing against each other. That is not as simple as it may sound. Points used in orienteering are usually the smallest points a human being can expect to find. A square foot or two in the middle of the woods, for example. Historically, orienteers were limited to a map and an old-fashioned compass to find their targets. Today, of course, orienteers use sophisticated GPS mapping devices. Orienteering began as a military exercise to train soldiers to find their way around unfamiliar territory. Today it is a popular sport. Find an orienteering club in your area and give it a try.

GIS technology changes constantly. Industries and organizations find new ways to use existing technology and create new systems along the way. Stay on top of these trends as they unfold. There are many websites covering all aspects of GIS. A few of them are Esri, GIS Lounge, and the Cartography and Geographic Information Society.

HISTORY OF THE CAREER

A SIMPLE MAP ON PAPER IS rudimentary GIS technology, and paper maps have been around for centuries. Of course, there is more to GIS than just getting from point A to point B. Essentially, GIS is the melding of computer science, statistical analysis and cartography. The advent of computer science is what made possible GIS in its modern form, but statistical analysis and cartography have been around for a very long time.

Statistical analysis and cartography were first combined in the 1800s to track cholera outbreaks. The science of tracking public health is called epidemiology, and is one of the most important uses for GIS technology, even today. In 1832 a French geographer named Charles Picquet created a map showing all 48 districts of Paris to illustrate the prevalence of cholera throughout the city. By using different colors to represent different percentages of cholera infection in each district, he created a simple map in which the health of each district could be determined at a glance.

Picquet did not, however, attempt to find the source of the cholera outbreaks. Connecting the dots was left to John Snow, an English epidemiologist who precisely mapped individual cases of cholera in London's Soho neighborhood. By connecting the dots on the map he was able to trace the source of the outbreak to a single contaminated water pump on Broad Street. He disconnected the handle and the cholera stopped. Snow did not have computers to assist him, but he did have maps and statistical analysis. By combining the two he figured out how to halt the spread of disease in a densely populated city. As simple as this now seems, it was a significant early example of the GIS thought process.

The development of photography and associated photographic films allowed cartographers to add layers to simple two-dimensional maps. Transparent film enabled cartographers to print a basic paper map and then add additional information like elevations, buildings, forested areas, geological formations, and many other features that needed to be mapped. Stacking a few transparent sheets on top of a map often revealed information that would not have been easily spotted before, such as the geographic distribution of buildings in relation to a fault line or some other geological feature of concern. This technique was widely used – at least as a backup to digital systems – until the early 2000s.

The first computerized GIS systems were developed in the 1960s. Various governments and military services created what would now be considered primitive mapping applications that allowed cartographers to add data to maps and search for data points and clusters in relatively small areas. The first true GIS system was developed by the Canadian government in 1960, and was known as the

Canada Geographic Information System. As the name implies, the CGIS stored data collected for the Canada Land Inventory, a government program to manage land use. Data entered into the CGIS could be manipulated and analyzed easily. It used overlays and national coordinates to enable users to compare data from different regions, making land-use analysis easier and much faster. The system was used by the Canadian government until the 1980s.

GIS technology grew alongside computer technology. Improvements in visualization technology allowed even complex information to be displayed on a computer screen for easier analysis. Better digital storage methods allowed more information to be easily available, and improvements in computing power enabled early GIS technicians to analyze more data in less time.

The 1970s and 1980s saw the creation of GIS systems with names like Odyssey, SyMap and GRID. The big leap came in 1986, when the Mapping Display and Analysis System – better known as MIDAS – was released as the first mapping system for the then-new Microsoft Disk Operating System, the standard for desktop computers. MIDAS eventually grew into MapInfo, a GIS company that still exists today. You can check out their products at www.mapinfo.com

Today's GIS products are both incredibly sophisticated and wonderfully easy to use. Industry-standard ArcGIS, for example, allows users to add almost any information they want to an enormous database of maps. This information is searchable and can be easily analyzed and interpreted. Maps can be sent from a head office to employees working in the field via tablets and smartphones, where they can be used, modified and sent further afield or back to the head office. ArcGIS can download data from thousands of sources, allowing users to add to their resources.

The military uses a variety of GIS systems, like FalconView, a variation on a popular GIS system anybody can access online.

Most GIS systems are custom-built, at least to some degree. Even off-the-shelf systems not subject to any special programming take on a custom-made nature after they have been used by an organization. Data specific to the organization starts to populate the database, users interested in certain types of analysis use the system in very specific ways, and over time the system becomes specialized in ways that its creators may not have anticipated. This is part of the challenge of a career in GIS. You can be a part of the process.

WHERE YOU WILL WORK

THERE IS NO GEOGRAPHIC CENTER OF the GIS industry. The whole idea of GIS is to be able to gather geographic data from anywhere in the world. There are, however, a few industries in which GIS is extremely common that could influence where you live and work.

The military uses GIS continuously. The US military has mapped much of the globe in minute detail using the latest in GIS technology. If US forces ever need to be deployed to a particular corner of the world, it is essential for them to have accurate detailed maps. The military also uses GIS technology for targeting, surveillance, patrol and other classified missions. All branches of the military offer training in GIS applications and opportunities to put that training to use. If you want to pursue a career in GIS you should seriously consider spending some time in military service. You will get training and experience, and you will become eligible for the GI Bill, which will help pay for college.

Civilian government is also a major user of GIS technology. Cities, in particular, use GIS technology to map their jurisdictions to keep track of things like zoning laws, and commercial and residential development. GIS is also helpful for planning and maintaining infrastructure like water and sewer lines. Police departments use GIS technology to track crime trends so they can put police where they are needed most. The US Census Bureau aggregates population data into GIS systems to generate statistical data about the population of the country. There are many GIS firms and in-house GIS departments in major metropolitan areas.

The nature of GIS means that many projects take place in exotic corners of the world. Mapping oil fields in Africa, for example, or urban development in Asia. Many GIS professionals become contractors who rotate from one project to the next, sometimes living at home and working in an office, sometimes living and working in the field, which could be anywhere.

YOUR WORK DUTIES

Geographers

Geographer is a very broad title that encompasses people working in government, for private businesses, the military and universities. Most geographers have earned master's degrees, and many hold a PhD degree, especially those working for universities. All geographers use GIS technology and most of them have had a hand in creating or customizing GIS systems, even if they spend most of their time as GIS users rather than creators.

When you are in school, geography is often reduced to basic knowledge like identifying US states or foreign countries on a map, or learning how to read map legends and find the way to get from one place to another.

Geography as a profession is about much more than simply knowing where things are. Geographers try to understand how things got where they are, why they persist, who is behind them, when they were created, and a host of questions related to their specific area of interest. They all use GIS technology to answer these questions.

All geographers start out with a similar set of functions. They gather geographic data through a variety of means, including field observation, satellite imagery, census and survey data, maps and photographs, among other tools. Then they use statistical analysis to find the answers they are looking for. Along the way they create and modify maps, prepare reports and advise people (including employers and clients) who need geographic data for specific purposes.

Historically, this process consisted of multiple separate steps that were synthesized by the expertise of the geographers doing the research. GIS technology combines the traditional functions of cartography and statistical analysis with the new technology of computer science to create a total picture.

There is a wide variety of work available for geographers. When they hear "geography," most people think of physical geographers who study the land, but there are also human geographers who study human activity and its relationship to the physical geography around it.

Regional geographers gather and analyze geographic data on a particular region, from a city block to an entire country or bloc of countries. Urban geographers home in on cities. Political geographers use geographic methods to track political structures and allegiances based on geographic factors like proximity. Environmental geographers study the natural environment and the impact that humans have on it, both positive and negative. There are many other specialties.

All geographers collaborate with GIS programmers to some degree. While many GIS systems start out as software applications, they are usually modified to meet the needs of specific users. It is very common for GIS companies to embed personnel in geographic enterprises, at least initially, to get to know how the product is used and what can be done to improve it. Users are in constant contact with creators so that patches and updates can be created as they are needed.

Sometimes there is no real distinction between users and creators. A geographer working at a university, for example, could spend time gathering and analyzing data for a project, and also serve on a development team creating new GIS technologies. Many universities are awarded public and private grants to develop new GIS technologies or work as contractors for government agencies like the Department of Defense. These are coveted positions for anyone who wishes to earn a doctoral degree and dig deeply into the promise and possibilities of GIS.

Geographic Information Specialists

Geographic information specialists are the technicians of the GIS field. Sometimes known as *photogrammetrists* or *cartographers*, geographic information specialists make maps and charts for a wide variety of uses. Very few geographic information specialists now use paper or acetate sheets to create maps, although rudimentary techniques are still widely taught as backups in case digital technology is not available.

Geographic information specialists often work for geographers, businesses, or universities that create GIS systems. They are also the largest category of end-users for GIS systems, and may spend time working with creators to design or modify GIS platforms. Geographic information specialists usually have bachelor's degrees, and many earn master's degrees after a few years on the job.

Geographic information specialists gather and analyze data and use it to create maps for specific purposes. Those purposes can be anything from animal migration patterns to human habitation, forest planting or deforestation, expansion or contraction of bodies of water, or the routing of roads and highways, for example.

GIS systems combine these cartographic functions with photogrammetry, which is the process of taking and aligning photographs to add detail to maps, usually through aerial surveys.

Geographic information specialists use many technologies to gather information and prepare maps. Geodetic surveys of the land are often used, as are images taken by aircraft and satellites Satellite imagery has improved dramatically in recent years. If you do not have

Google Earth installed on your computer already, you should do it right now. Link is at: www.google.com/intl/en/earth/index.html

Geographic information specialists also use a technology called LiDAR, which is short for "light detection and ranging." LiDAR systems use a laser mounted on an airplane to sweep the earth with pulses of light that are reflected back to the airplane and then processed through a powerful computer to create a three-dimensional image. By using the tiniest trace of reflected light LiDAR systems can sometimes see inside buildings and under foliage. They can produce bare-earth models that strip away clutter to reveal the contours of the earth below the trees and manmade structures. LiDAR has become a favored tool for military uses and for physical geographers who need to see through things on the ground.

GIS Specialists in the Military

All of the military services use GIS professionals to accomplish their missions. Nobody trains and employs as many GIS professionals as the US military. Many GIS professionals get their start in the military and then move into private sector employment. In one five-year hitch they can get top-notch GIS training, demanding experience in the field, and qualify for the GI Bill to earn a college degree after they leave the service.

All of the services use GIS technology in slightly different ways. They all need maps, however, and they all need to figure out how to target their weapons. GIS is an important part of the military function known as C4ISR, which is an acronym that stands for "command, control, communications, computers, intelligence, surveillance, reconnaissance."

Mapping is so important to defense that the Department of Defense maintains several agencies devoted solely to cartography. The National Geospatial-Intelligence Agency, or NGA, collects, stores, and distributes geospatial intelligence to all branches of the armed services. NGA employs several thousand geospatial experts at its headquarters in Virginia and on military bases around the world.

The Defense Mapping Agency coordinates all mapping functions for the Department of Defense, including printing and stockpiling paper and digital maps in a variety of classifications. The local DMA office is often the first stop for military personnel preparing for a mission.

All of the services maintain their own geospatial functions according to their individual needs, such as maritime charts for the Navy, ground maps for the Army, and aeronautical charts for the Air Force.

The military uses GIS technology in two basic ways: to get from one place to another, and to target weaponry. Ships and aircraft have sophisticated systems to guide them. Both ships and aircraft navigate with help from the Global Positioning System, a constellation of satellites that supplies location data to users on the ground, at sea, and in the air. Soldiers on the ground also use GIS data when on patrol using what are essentially military-grade smartphones. Portable mapping devices have proven very adept at helping planners to put the right amount of force in the right place at the right time, and to help soldiers and Marines in the field to accomplish their missions.

The military also uses GIS technology for targeting. Targeting is not only the process of aiming a weapon, but also the process of determining the whereabouts of a target to aim a weapon at. The military uses satellite and airborne surveillance, LiDAR, electro-optical systems, and many other technologies to gather and analyze information. All of these systems require GIS expertise.

GIS Software Developers

Behind every computer application is a software developer. For some GIS professionals, it is the computer science that matters most.

GIS developers work in a wide variety of environments. Many are employed by universities and develop GIS platforms under contract to a government agency like the Department of Defense. Others work for businesses in the private sector, and create GIS tools that can be applied to many functions.

GIS developers typically have degrees in GIS or computer science, and often graduate degrees up to and including a PhD. They may or may not have set out to specialize in GIS, but they know their way around software development.

GIS PROFESSIONALS TELL THEIR OWN CAREER STORIES

I Am a Geographer for the Army Corps of Engineers

"When people hear the name Army Corps of Engineers, they immediately think of legions of soldiers building things. The Army Corps of Engineers — commonly known as ACE — is actually the federal agency responsible for maintaining much of the nation's infrastructure. We own and operate more than 600 dams, for example, operate and service 12,000 miles of inland waterways, supervise nearly 1,000 harbors, and regulate the use and maintenance of the nation's wetlands. We also support Army and Air Force installations and conduct military construction projects around the world, but most of our employees are civilians engaged in civilian projects.

I was always interested in geography as a kid. Memorizing the states and all of the countries of the world wasn't enough for me. I had to dig into the details, figuring out the lay of the land, literally, and how it affected everything on it. I earned a bachelor's degree in geography and got my first job with a state waterway-management agency. In that job I used GIS to keep track of a small system of inland lakes and rivers, including flooding, the state of forests within the watershed, the depth of harbors, and the use of land near the waterways. I loved it but wanted to move up in the world. I earned a master's degree in geography and landed my current (civilian) job with the ACE.

I get to use all the latest technology in this job. I can pull up-to-the-minute satellite imagery covering any spot in the United States, and in many of the places where we have significant interests. I can use LiDAR imagery to create layered depictions of the earth and everything on it in a given area. I also use computer-aided design (CAD) tools to render geographic features, and global positioning system (GPS) tools to precisely locate points on the earth. I handle photogrammetry from electro-optical imagery, and manage several relational databases.

My job really underscores how closely GIS creators and users work together. I generate three-dimensional terrain models and perform land surveys, but I am also responsible for working directly with managers, engineers and scientists to develop new GIS applications. I am in charge of leading the plans for implementing new GIS technology within my office. Technology evolves very quickly, and every time there's a big leap in capability, people like me dream up hundreds of new uses for it.

I'd recommend this career to anybody who is as passionate about technology as they are about geography."

I Am a Military Targeteer

"Targeteer" is what the military calls personnel assigned to find targets, pinpoint coordinates and conduct battle-damage assessments after a strike. This job is typically assigned to intelligence officers, but in a forward-deployed environment, it could go to enlisted intelligence specialists or to personnel in various aviation communities.

We call targeting the science of 'putting warheads on foreheads.' A bit crude but basically accurate, at least as far as it goes. We target lots of things – buildings, enemy tanks and artillery, ships, aircraft, and small targets like installations hidden in the woods. We use all manner of sophisticated GIS technologies to locate targets, some of which I can't talk about, but some I can.

We use electro-optical imagery, or EO. EO is basically conventional digital photography enhanced by specialized lenses and digital imaging devices. We can take thousands of pictures of a given territory and then put them together in the computer to create a huge mosaic that we can search and analyze. This is known as photogrammetry. Old-timers tell me that they used to cut and arrange actual printed photographs to build their mosaics. GIS sure makes this a lot easier.

We also use LiDAR, which stands for 'light detection and ranging.' LiDAR cameras are mounted on airplanes and use a laser to scan the earth. The cameras pick up returns when the laser bounces back. The speed and intensity at which the light returns is translated into a detailed topographical map of the ground below. The system can tell the difference between water and solid ground, for example, or between foliage and buildings hiding beneath it. I can flip a LiDAR image upside-down and look into holes in the surface.

I have a plan for my GIS career. I've been in the military for about three years and have two to go. I have had some of the best GIS training the world has to offer and plenty of opportunity to put my training to use in demanding environments. After I get out, I intend to cash in my GI Bill to earn a bachelor's degree in GIS. After that, I think I'll look into contracting for the Department of Defense. A lot of the people doing GIS work in forward-deployed locations are actually civilian contractors working for the Department of Defense on relatively short-term contracts of three to 12 months. They are paid very well, too.

If you want to pursue a career in GIS don't overlook a hitch in the military. Nobody trains more GIS professionals than the United States military and it's hard to beat the experience. You may not think the military is for you but if you give it a try you may be surprised."

I Am a GIS Software Developer at a University

"I didn't set out to get into a GIS career. I majored in computer engineering and earned both bachelor's and master's degrees in the subject straight out of high school.

Computer engineering was a good fit for me because I enjoy the combination of hardware and software, of making computers and associated peripherals that work together to do amazing things. Little did I know what waited for me a few years down the road.

After graduate school, I went to a job fair for techies like me and chatted up a recruiter from a science lab at a major university. I quickly discovered that we spoke the same language, so to speak, and seemed to have a lot in common. A few weeks later I was hired and started my new job.

At first, the lab put me on several large projects, working alongside seasoned veterans to get to know the lab and how it did its business. After a couple of years I was assigned to a small project to design a new LiDAR system that would dramatically increase coverage and accuracy. The new system could be used by surveyors, forestry agencies, local and state governments and the military. The focus of the project was on developing new capabilities for the mechanical elements of the LiDAR camera, and then integrating those new capabilities into GIS software.

We tested out our new device by flying it over the jungles of the
Philippines to see if we could pick out manmade structures beneath the forest canopy. This is not easy to do. Our system generated millions of individual points from the laser returns off the foliage and, occasionally, the forest floor. When we got enough hits off the ground, we could put together an image in the computer and effectively see through the thick canopy of foliage.

We patented our new LiDAR system and have since contracted our services to many users. As one of the leaders of the program I have been able to spend weeks or months at a time all over the United States, and in the Philippines, Brazil, Guatemala, and Afghanistan, all of which were amazing adventures. GIS careers are incredibly complex and use a wider variety of skills than you might think. I didn't even know what GIS was until I got this job, and now I'm at the leading edge of the technology."

I Am a Geographic Information Specialist for a Forestry Agency

"Want to know how many acres are covered by trees in the northeastern corner of the state? And what species they are? Or how about elk migration patterns through state parks? I can answer all of these questions, and more.

I ended up in GIS after working as a surveying technician in college. I went to college to major in business administration, but switched to GIS because I decided I liked my part-time job assisting surveyors with plotting property lines and such. Never saw that coming, but I'm glad it worked out the way it did. Earning a bachelor's degree in GIS is what made me competitive for my current job with the forestry agency.

Here's the thing about GIS. Sometimes it's hard to tell if the job is about GIS or about the subject you use the GIS tools for. I came to the forestry agency because of my background in GIS, but a lot of the people here have degrees in forestry and earned professional certificates in GIS after they had been here for a few years. You can set out to become a forest ranger and become a GIS expert along the way. The same could be said of many other professions that use GIS as a critical tool.

The thing that makes us all GIS experts is the fact that most GIS systems are custom-built to serve each customer's needs. Those needs evolve over time. So does GIS technology. Most GIS users have long-term contracts with their system designers to get frequent updates, and to make sure that new capabilities are incorporated into the system. So while you may be mostly a user of GIS technology, you're never very far removed from the creators who manage your system. They may be sitting next to you, asking you questions and giving you updates. That's certainly the case in my agency. We use a version of ArcGIS modified for our needs. The company that makes the software is in constant contact with us to make sure we are getting what we need.

We use data from a wide variety of sources to do our work. The federal government maintains enormous databases of satellite imagery from which we can pull images for analysis. We also use imagery from commercial sources and gather our own, mostly via electro-optical and LiDAR cameras mounted on aircraft.

We can crunch numbers on just about anything. We track deforestation on the fringes of urban areas and state parks, for example. We can also follow wildlife migrations. By using LiDAR bare-earth models we can even track the very slow erosion of topographical features like hills and riverbeds. All of these data are important environmental markers that allow us to monitor the health of the ecosystems for which we are responsible. Without GIS I'm not sure how we would do it. Walk around the forests and count the elk? They run pretty fast, you know."

PERSONAL QUALIFICATIONS

THE CAREER REQUIRES A VERY HIGH degree of mathematical ability, whether you are a creator formulating complex algorithms, or a basic user collecting and interpreting data. Geometry is used to calculate everything that makes GIS systems so much better than two-dimensional maps, including topography, distance, and the size of manmade features on the ground. While it is true that the GIS system will do most of the heavy number-crunching for you, you will not be able to interpret what you are looking at unless you have a pretty sophisticated understanding of the math behind the calculations.

A GIS system can churn out an avalanche of data in a short time, but it is the responsibility of the user to determine what the data mean. Using a GIS system to calculate the number and average size of buildings within a particular area is impressive, but the results are meaningless unless they serve some greater purpose. What questions can be answered with this knowledge? Knowing the number and size of buildings can give city planners a good idea of how big sewer lines need to be, or how many classrooms a new school will need. Military planners can use the same data to determine where the enemy may be hiding, or how many blankets and bottles of water they will need to distribute to disaster victims. Numbers only mean something when they can be put to use. That will be your job.

This work requires vision and imagination. The most successful GIS professionals are those who can take the long view and then figure out how to get there. It is easy to envision a GIS system that will put all the data the user needs in one place, in an easy-to-interpret format, from a smart phone and in 3D! Like most computer-based applications, GIS systems tend to move forward incrementally, with one little breakthrough leading to the next. Somebody has to know what the final product needs to do. Scientists and engineers often refer to this process as "working backward" from an ideal end goal, and figuring out how to fill the gap between the goal and the reality.

ATTRACTIVE FEATURES

GIS IS A FASCINATING CAREER, ON THE cutting edge of technology. The military doesn't target missiles with pocket calculators. Land-use planners don't calculate flood plains with tape measures. The Coast Guard doesn't plot search-and-rescue missions with guesstimates. GIS systems are essential to all of these complex missions, and countless more. GIS professionals utilize computer science, statistical analysis, and cartography in designing and implementing some of the world's most complicated and productive systems. The first time you see an analyst flip a digital topographical map upside-down so you can see what is under the ground you will be impressed. You can be a part of this amazing world.

It is hard to overstate the importance of GIS to modern life. GIS affects much more than just the professions who use maps and basic geographic information. Banks use spatial models generated by GIS data to track economic conditions and serve their markets better. Journalists use GIS to conduct research and create graphics for use in print and online. GIS is used to study the ocean floor, and monitor the health of the world's forests. GIS can even analyze features below the surface of the earth, giving mining companies an edge in finding and exploiting natural resources. You will be using your skills to accomplish something important and far-reaching that affects the lives of millions of people.

Advances in GIS have made the technology more useful and less expensive, meaning GIS finds its way into more sectors of the economy every day. GIS job opportunities are expected to grow about 20 percent faster than the average for all professions over the next 20 years. This healthy growth in demand will create new opportunities for you in the form of new jobs, new technologies, and new educational options.

UNATTRACTIVE FEATURES

THE FIELD IS EXTREMELY competitive The GIS career attracts extremely intelligent and highly motivated careerists who want to make their mark. Where there is room for visionaries and pioneers, there will be extraordinary people seizing the moment. Look at the competition as an opportunity to work hard and rise to the top. Even if you are willing to go with the flow from the middle of the pack, you can be sure you will be surrounded by fascinating people making the world a better place. If your goal is to succeed at the highest level, you will have to work very, very hard.

Most GIS systems are customized for specific users and their needs. GIS professionals spend endless hours working together to design systems and to fine-tune them over time. You have undoubtedly seen software applications labeled according to their latest update: Version 1.3, 2.4, etc. In GIS, it is not uncommon to find applications with labels like Version 24.396 or simply

Version 427. This is because GIS technology is constantly evolving, and GIS users are always finding new ways to use it. Even the best custom-made system will not last more than a few months without an update.

Ask any psychologist, and they will tell you that most people do not like change. People may think they like change, and they will sing the praises of enterprising people and organizations that lead change and make the world a better place, but most people are happier when change happens slowly and does not disrupt their daily lives. Visionaries embrace change and figure out how to master it. This relentless competition is not for the faint of heart. Change in the GIS business is both rapid and necessary, and it will be up to you to keep up. Your employer will not tolerate dead weight.

EDUCATION AND TRAINING

NOT SO LONG AGO YOU COULD almost stumble into a career in GIS by earning a bachelor's degree in geography, statistical analysis, or computer science, or by getting very specific training in GIS through some other means, like a hitch in the military. To some extent this is still true, but colleges and universities have responded to the high demand for GIS specialists by offering a very large array of degrees and certificates in every GIS specialty imaginable.

The simplest path to a GIS career is to earn a bachelor's degree in the subject. Often known as geographic information science or geographic sciences, a bachelor's degree in GIS will give you a broad foundation in the elements of GIS. Typical courses include human geography, spatial thinking, geospatial tools, physical geography and statistics. Some programs allow students to declare a concentration, such as conservation, urban planning or remote sensing. Many conventional geography programs offer concentrations in GIS. GIS can also be a minor or double major taken in conjunction with a program in computer science or statistical analysis.

You may decide to earn a master's degree in GIS. There are many options available, which allow further specialization in the area of GIS that interests you most. Part-time programs for working adults are quite common. Most careerists do not really focus on the part of GIS they like best until they have been out in the field for a few years. You may need a master's degree later in your career, but find out what you like best before continuing your education past college.

You should not miss the opportunity to complete an internship. A full-time job related to your major that takes the place of classes for a summer or semester, most internships are paid and they all come with a once-in-a-lifetime opportunity to try on a career for a few months and walk away without burning bridges behind you. As an intern you will work alongside your full-time counterparts in GIS and have a closeup view of what they do in a typical day. There is no better way to learn how a career really works than by completing an internship. It is common for companies to hire interns in their first full-time jobs after they graduate.

There are also many certificate programs available at both the bachelor's and master's degree level. Certificate programs tend to be aimed at working professionals who have some professional experience, even though they may or may not have relevant academic degrees. Certificates are also appropriate for military veterans who have extensive training but have not necessarily earned a degree.

There is no better GIS education than that provided by the United States military. All of the services use GIS technology to find their way around the land, sea and air. GIS is especially important for creating the precision maps that are used for targeting weapons and directing infantry and special forces teams. The military constantly updates its GIS databases so they will be ready when they are needed. GIS professionals – the services all have different names for their GIS personnel, as well as different training pipelines – work on databases when they are at peace and make sure that deployed forces have the GIS products they need when they are at war. Depending upon the service you choose, you could enlist for as few as four years and return to civilian life with some of the best GIS training and experience the world has to offer. Then you can cash in your GI Bill and earn a degree.

EARNINGS

EARNINGS FOR GIS CAREERISTS VARY considerably depending upon the exact nature of each career path. Federal government employees, for example, are paid according to the General Schedule, a system that grades most federal employees from GS-1 to GS-15, with 10 additional steps within each pay grade. There are too many federal jobs involving GIS to list here, but GIS careerists are likely to start their federal careers at the grade of GS-9. Pay for GS-9 employees currently ranges from about $41,500 to $54,000 annually. Pay for grade GS-15, the top of the GS pay scale, ranges from over $99,000 to about $129,000. All federal government employees also receive a standard package of benefits including health insurance and paid vacation time. State and city employees are paid in similar fashion, according to their own scale.

Military personnel are paid according to a similar scale. Enlisted personnel at the rank of E-1 are paid about $1,500 per month, not including the value of housing and meals. After three to four years, most enlisted personnel will rise to the rank of E-4 and earn over $2,300 per month, and be eligible for a housing allowance so they can live off base. Enlisted personnel at the rank of E-6 are paid almost $3,200 per month, in addition to a housing allowance.

Officers are paid more, but also have greater responsibility and need to possess a bachelor's degree in order to apply for a commission. All military personnel are also eligible for special pay for deploying to combat zones, or for demonstrating a valuable skill like a foreign language.

GIS professionals working for universities or corporations that create GIS systems can be paid anywhere from $60,000 per year to more than $200,000, depending upon their employer and the type of projects they work on.

Contractors working for the Department of Defense on short-term contracts can be paid as much as $80 per hour while they are deployed, which is almost $6,000 for a 72-hour workweek.

OPPORTUNITIES

THERE ARE SEVERAL WAYS TO MOVE up in the world of GIS. Because the technology is constantly changing and evolving, there are always additional certificates to be earned. Do an online search for "GIS certificate" and you will get hundreds of hits. Educational institutions from prestigious universities to small, online-only training firms offer certificates and other credentials in every possible specialty in GIS. If you pursue this career, you will have to return repeatedly to the classroom (or online resource) to learn the latest developments in the field. The nature of GIS lends itself to online learning, so earning new certificates does not have to involve a classroom schedule.

There are innumerable opportunities to use your GIS skills. The Department of Defense, for example, hires thousands of GIS specialists every year to take on short-term contracts supporting military deployments around the world. Contractors typically spend anywhere from three to 12 months deployed alongside uniformed military personnel, and are responsible for supplying the GIS data needed by forces in the field. As civilians, contractors are rarely required to leave the safety of major bases (and may even be prohibited from doing so), and they are not required to carry weapons. They may live in tents, however, and are often required to work at least 72 hours per week. It is not easy work, but it pays extremely well, gives you experience that will look great on your résumé, and is nothing if not exciting. When your contract is up, you get to go home.

The GIS business rewards visionaries. There is no such thing as a "finished" GIS system. GIS platforms are constantly updated and reconfigured to keep up with technology and meet the needs of customers. GIS systems are rarely supplied to customers "as-is." They are usually modified and customized to meet specific needs. Even if you are mostly an end-user, you will be expected to innovate and to push the boundaries of your system and what it can do. Those who can see past the present and find ways to make their systems the very best they can be, will be rewarded.

GETTING STARTED

YOU SHOULD BE READY TO TAKE YOUR first steps into the real world even before you graduate from college. Get your personal marketing materials in order. A résumé typically gets about 10 seconds of scrutiny before it goes into the trash, or into the much smaller pile to be called for an interview. Today, many résumés are electronically read by a computer looking for keywords related to the job. You can bet the computer does not even take a whole 10 seconds to read your résumé. You need to grab the reader's attention, whether that reader is human or electronic. There are many books and software applications that can help you put together a first-class résumé. If you are not confident in writing skills do not hesitate to pay somebody to help you. Even though many job applications are online, take the time to prepare a résumé properly formatted in the traditional manner. You may be asked to email a résumé to somebody who will print it out on their end. A polished résumé also makes electronic forms easier and more accurate by giving you a ready source from which to cut and paste information. Take a hard look at the description of the job you are applying for and feel free to sprinkle a few keywords from the description around your résumé. Sometimes using the exact word can make the difference between the computer choosing your résumé or skipping over it.

Start your search by getting in touch with the people you have met in the GIS business. Your studies and an internship should have provided you with a stack of business cards and email addresses. You would be surprised at how many recent graduates get their first real jobs at the same business that hired them for an internship. You are a known quantity, already familiar with the company and its culture. If you were a good fit for an internship, you will probably be successful in a long-term job. There is no guarantee that your old employer will have an opening, however, so be prepared to look further afield. Many of your contacts will have their own contacts they can refer to you.

More than anything, do not give up. Your first job does not have to be your ultimate dream job. It just has to get you into the business, where you can start learning and make connections. Do not hesitate to take the first good opportunity that comes your way. It is only the first rung on the ladder to the career of your dreams.

ASSOCIATIONS
PERIODICALS
WEBSITES

ArcGIS
http://www.arcgis.com/features

Boston University
www.bu.edu

Bureau of the Census
www.census.gov

California State University Los Angeles
www.calstatela.edu

Cartography and Geographic Information Society
www.cartogis.org

Central Michigan University
www.cmich.edu

Directions Magazine
www.directionsmag.com

Doyle's GIS Links
www.doylesdartden.com/gis

Earth 3D
www.earth3d.org

Eastern Illinois University
www.eiu.edu

Esri
www.esri.com

FalconView
http://www.falconview.org/trac/FalconView

Federal Geographic Data Committee
www.fgdc.gov

Geo Community
www.geocomm.com

Geographic Information System Learning Site
www.ccdmd.qc.ca/en/gis

George Mason University
www.gmu.edu

Geospatial Information and Technology Association
www.gita.org

Geospatial Media
www.geospatialmedia.net

GIS Lounge
www.gislounge.com

Google Earth
www.google.com/earth/index.html

International Cartographic Association
www.icaci.org

Kansas State University
www.k-state.edu

MapInfo
www.mapinfo.com

Marble
www.marble.kde.org

Massachusetts Institute of Technology
www.mit.edu

NASA World Wind
www.worldwind.arc.nasa.gov

National Geospatial-Intelligence Agency
www1.nga.mil

National States Geographic Information Council
www.nsgic.org

Notre Dame University
www.ndu.edu

Open Street Map
www.openstreetmap.org

Open Geospatial Consortium
www.opengeospatial.org

Open Source Geospatial Foundation
www.osgeo.org

Open Web Globe
www.openwebglobe.org

Participatory Geographic Information Systems and
Technologies
www.ppgis.net

Penn State University
www.psu.edu

San Diego State University
www.sdsu.edu

United States Air Force
www.airforce.com

United States Army
www.goarmy.com

United States Army Corps of Engineers Army Geospatial
Center
www.agc.army.mil

United States Coast Guard
www.gocoastguard.com

United States Geological Survey
www.usgs.gov

United States Marine Corps
www.marines.com

United States Navy
www.navy.com

University of Arizona
www.arizona.edu

University of Illinois
www.illinois.edu

University of Wisconsin
www.wisc.edu

West Virginia University
www.wvu.edu

World Wide Telescope
www.worldwidetelescope.org

Copyright 2015

Institute For Career Research

Website www.careers-internet.org

For information on other Careers Reports please contact

service@careers-internet.org

www.ingramcontent.com/pod-product-compliance
Lightning Source LLC
Chambersburg PA
CBHW070748180526
45168CB00004B/1564